Riding the Blue Marble

A Meditation on Faith, Science, and Hope

Dave Wasserman

Publisher: BookBaby

Imprint: Crooked Creek Publishing

ISBN: 978-1-54392-688-0

INTRODUCTION

This illustrated essay is intended to be read slowly. A meditation about our planet needs time for reflection and consideration and conversation.

The images of our Planet Earth come from the United States Government's National Oceanic and Atmospheric Administration (NOAA). Science on a Sphere® (SOS) is a research and educational resource that uses satellite-acquired information and mathematical models to report and project trends about Earth's various systems.

The text offers a spiritual overlay to Earth's story. The images reflect the beauty of the planet upon which we reside.

All this is offered in hope. Hope that some familiar ideas presented in a slightly new way may stir us to become more thoughtful caretakers of God's garden.

Dave Wasserman

April, 2018

DEDICATION

This meditation is dedicated to the youngest generation of Earth's peoples, the ones who may live to see the year 2100. Among the children around the world are those in our family and lives: Addie, Tali, Dylan, Jackson, Asher, Julian, Mary, Harvey, Pearl, Charlie, Henry, Blake, Aidan, Isa, Jackson, Wyatt, Jack, Will, Owen, Molly, Megan, Mazie, Piper, Wren, Max, Henry, George, Charley Jane, Mandy Kate, Meredith, Chase, Sophie, Jace, Avi, Jack, Nathan, Noah, Grace, Natalia, Evie, Alex, Oliver, and Steven.

May we in our time become better stewards of the resources entrusted to us, for their sakes.

For the beauty of the earth,
for the glory of the skies,
for the love which from our birth
over and around us lies,

Lord of all, to thee we raise
this our hymn of grateful praise.

For the beauty of each hour
of the day and of the night,
hill and vale and tree and flow'r,
sun and moon and stars of light,

Lord of all, to thee we raise
this our hymn of grateful praise.

Hymn Text: Folliott S. Pierpoint (1864)

in the beginning...

at a 4:00 am on-the-open-ocean darkness
in a polar freeze cold beyond cold
and the emptiest of grand canyons

the universe began
with a sudden explosion
and
a slowly rising dawn tip-toeing above the horizon

as if the Creator God pushed the Big Bang button
to set things in slow motion

in a few seconds, the very first particles formed
and then
quietly collected themselves
over 200 million years
in the silence and the darkness

until
that day-of-Creation sentence was uttered and
the first stars showered us
with glorious Light

and now,
over 13 billion years later
we find ourselves traveling
on a constantly moving planet

...spinning 24 thousand miles on earth's axis
...traveling 1.6 million miles around our starry sun
...rotating some 12.4 million miles as a member of
our galaxy's club of stars and planets and moons

every day
day after day

not quite lost in the space of a universe so big
that whenever we pause
for even the briefest of moments
to reflect
our souls can be stirred

so moved, we look around our planet
and recognize a Creator -
Someone with cosmic imagination
who started all this
and gave us a round and robust space-ship
setting us on this journey of

Life

Infrared image of our galaxy and its 200-250 billion stars: The white band is where the stars lie (including our Sun), the green and red specks are space dust, the dim white specks above and below are other galaxies.

from the moment our ancestors first stood up
we've looked to the heavens
searched the seas
mastered the mountains
and dug down deep within
thinking, reflecting, praying

hoping that by learning what we can about our traveling planet
we might discover our place and our part
in this journey

we've observed how events happen
considered how things work
and
studied and tested and argued and wondered

and come to realize...

this Earth is a
most precious place in the universe

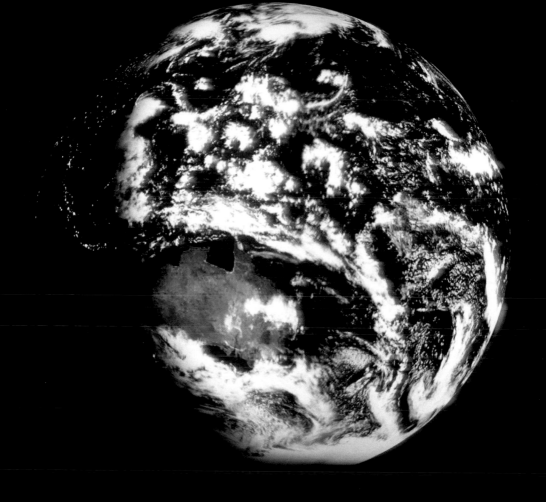

The Blue Marble - Earth viewed from space, including cloud cover:
This view is of the Australian continent. On the left is the edge
between night and day. The white dots are the nighttime lights of
Southeast Asia.

we know Earth comes packaged
with an inner core of fire, heat, melting rock

a crust upon which we walk
and seismic plates under the land and the seas
moving all the time
as much as an inch a year

colliding and separating
separating and colliding

the mountains rise up
the earthquakes rumble and jolt
the volcanoes blow off their steam
the tsunami waves erupt from the ocean deep

all this we know

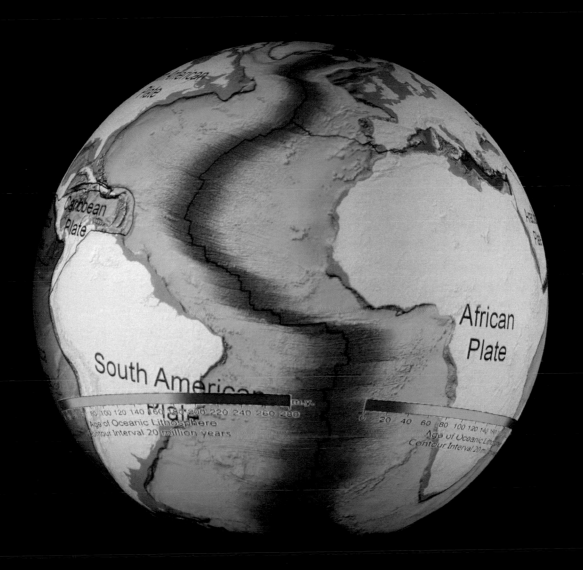

View of tectonic plates: Red and orange are younger plates; purple, above Africa in the eastern Mediterranean Sea, is the oldest plate on Earth. The black line (with red) down the center of the Atlantic Ocean locates the underwater mountain range from the North Pole to the South Pole (the longest range on Earth).

and

from our big-eyed telescopes and remote robots
we know our Earth has more water
than any other planet we've yet discovered

70% of the Earth's surface is covered in water
seawater in which our fish breathe - eat - swim
always-moving oceans
holding 97.5% of our planet's H2O
every drop undrinkable

we know too that 98% of the other 2.5%
is hard to reach
hidden in deep underground reservoirs
and frozen away in polar ice caps and glaciers

two percent of two-and-one-half percent:
not very much water for Earth's animals and
7.6 billion humans to drink
and use to wash ourselves, our clothes, our cars
to irrigate our corn, lima beans, onions, chiles
to water our forests and savannahs
to grow the grains that feed the animals for our diets

and in case you haven't heard the word:
more people are coming to dinner soon

View of currents in the Atlantic Ocean: North of the equator, the currents move clockwise. The Gulf Stream is on the left. South of the equator, the currents move counter-clockwise.

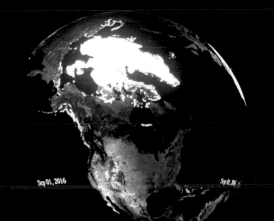

Sea ice concentration over the North Pole, September 2016: Ice grows and shrinks annually with the seasons, but the longer trend is that the total amount of ice is shrinking.

we know that
our traveling-through-the-stars Earth
has a very thin atmosphere that contains the air
we need to breathe

we know about the cycle of inhaling and exhaling
of oxygen and carbon
of our partner plants, trees, and forests
absorbing carbons, releasing oxygen
to keep us alive and lively

we know about the nitrogen that our plants themselves
must have to live

we know that this thin band of atmosphere
only 62 miles from surface to space
filters the life-giving energy our Sun sends us

we know about the Sun's light and heat and
ultra-violet radiation, the good and the bad,
and how this thin band of layered atmosphere
keeps the Earth warm enough to sustain life
but not so warm as to destroy it

and we know more...

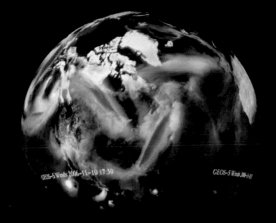

Jet Stream, November 2006, circling the North Pole (white) and moving over North America: As air passes closer to the Equator it warms, and then cools as it moves towards the poles. Colors indicate wind speed; orange is fastest.

Ozone hole (red) located over Antarctica, September 2017.

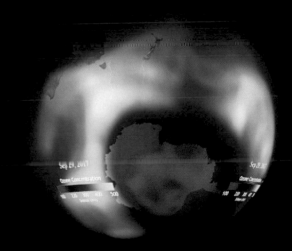

we live because
the Earth was made to balance

   heat and cold
     light and darkness
       wet and dry
         exertion and rest

we know that small changes can upset the Earth's
equilibrium

and we know that the Earth itself
as a part of its nature
has resilience -
that capacity to overcome some imbalances
returning land, water, air to viability

the ice ages have come and gone,
they may come again
lightning's fires burn and destroy
floods rise and recede
making way for new growth to appear

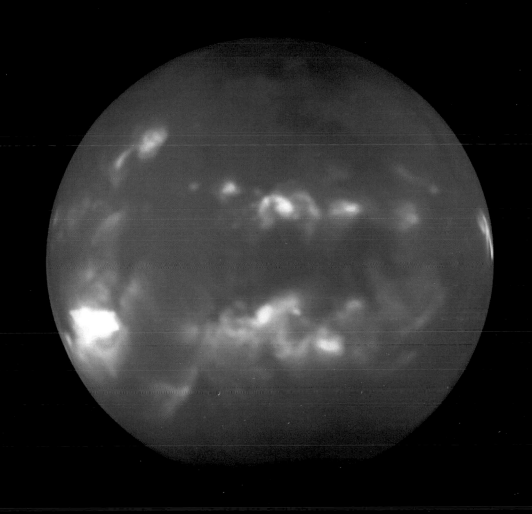

An X-Ray of our Sun (2003), source of heat, light, and infrared energy: Earth's atmosphere is designed to filter this energy to keep the planet's surface temperature within habitable tolerances. As the Earth spins and travels, the Sun sustains life's rhythms between day and night, light and dark, warm and cold, summer and winter.

we know much
about the way this old planet was formed,
how it works, and
how it makes Life possible with its balancing acts

the One who created all this
and has entrusted Earth to human beings
must be a loving One to have done so
must be One more interested in creating than destroying
must be One who decided that the living and the dying
of each one of us has purpose in the big scheme
— the universe — of things

perhaps that is why
for those among us on a Godly path
every breath we breathe
every drop of water we swallow
every moment of warmth in the Sun
is cause to say
Thank You. Thank You. Thank You.

this is the Garden
with all of its moving and balancing
handed over to us
to continue Life's journey

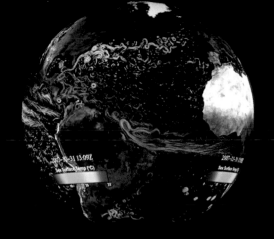

Ocean currents, part of Earth's heat management system: red/orange currents near the Equator are warmer; blue/green near the Poles are cooler. Warm water moves toward the Poles and is cooled before returning to the Equator.

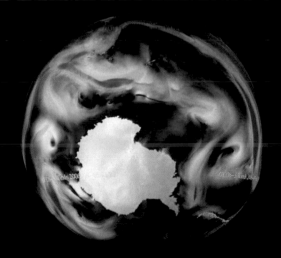

Jet Stream winds circling Antarctica, September 2006: Colors indicate wind speed, red/orange is fastest.

but there's a problem:

as Earth walks its balance beam
we creatures have not been paying enough attention
to our planet's groans and moans

we humans have been given the freedom to
organize life
to explore, discover, experiment
build, invent, manufacture, modernize
connect ourselves across the globe

freedom to fish and farm
build and burn
buy, sell and trade
dig and unearth
use and renew

freedom to divide ourselves into saints and sinners
followers of God or not

we are free

View of human transportation: The red lines are air travel corridors, the blue lines are shipping lanes.

How we use the grains we grow (corn, rice, wheat): Green is for human use, red is animal feed, orange is shared by humans and animals.

and

we are responsible
to pay attention to what we do and don't do
with our freedoms

responsible to live
with those-who-will-come-after-us in mind
and not simply ourselves
responsible to use but not over-use
the resources entrusted to us
responsible to pay attention to Earth's equilibrium
to be good and Godly stewards of our planet

if for no other reason...
because
Earth is our only home

so we know to pay attention to
the carbon we put in the air
because too much carbon overheats the planet
and the exhaust from our cars and manufacturing plants
doesn't disappear when it's out of sight -
it drifts and hangs out in our atmosphere

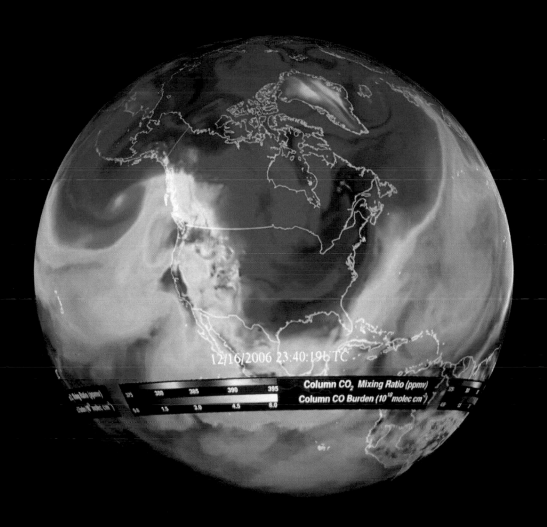

Carbon Dioxide emissions over North America, December 2006: Red and orange indicate higher concentrations (particles per million). A wispy image indicates where winds are moving the CO2 away from the source. Carbon does not leave the atmosphere quickly. North America is located under the orange area; South America is located in the lower right, below the map legend.

and
we know that when we turn on our lights,
fly our airplanes, charge our cellphones
we put more carbon into the air

we know that super-sizing
those concrete jungles we call cities
puts more heat into the air

we know that when carbon dioxide drops into the ocean
acid is formed that
destroys the coral reefs that
feed the fish and protect our shorelines

we know to pay attention to
how much we can trash Mother Earth
her land and her seas
without destroying our plant and animal life
and upsetting the symmetry our planet must maintain

still, we start our engines and toss our water bottles
and then act surprised
when the seas warm, the reefs die,
the air becomes un-breathable
and Earth's systems start to lean and tip
toward the uninhabitable danger zone

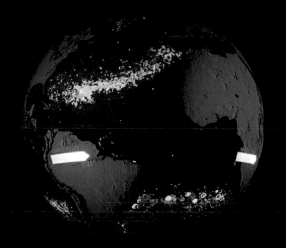

What if you tossed floating trash all over the oceans? Where would it go? One study shows our garbage collecting in the center of the circulating currents. This image (gray locates the continents) shows the partial collection accumulating in the Atlantic Ocean.

Ocean coral at risk in the Caribbean Sea: Darker colors indicate greater risk. Coral is also at severe risk in the waters north of Australia.

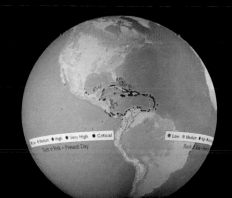

now...

increasingly over the last 50 years
those who study the Earth are

 ...telling us precisely what is happening
...sharing what will likely happen in the future
...and offering clues about when our planetary home
could reach its tipping points

but we are free
to ignore them if we choose...

because it goes against what we want
for ourselves or our generation
because it is objectionable to our way of thinking
because it is inconvenient to our way of living
because it is unfair that we are told to change
while others elsewhere don't
because life is really about ME and MINE

...or so we think

Volcano locations (green triangles) both above and below sea level along the Pacific Rim: from Australia (lower left) to the Philippines, Japan, Alaska, the US and Central America.

Mapping human violence (red) and human protests (purple), based on newspaper reporting around the world: The greater the number of newspaper articles, the larger the circle. This view was taken on January 22, 2017, over the Western Hemisphere.

so where is the Holy One in all of this?
       Creation, Fall, Redemption

...the Creation, we religious folks say, is good
...the Earth's fall, we suspect, is dangerously possible
...and the redemption?

redemption comes through our sacred teachings and wisdom,
through the bold acts of mercy and sacrifice
of our spiritual saviors

redemption comes by acknowledging
that Earth's stability must be protected
if Life is to be sustained
maybe our scientists are God's angels
bringing messages of warning and hope

maybe redemption comes because
good old Mother Earth still has resilience
and can regenerate the land, water, and air
when we work patiently with her

and maybe redemption comes
because each one of us has the capacity to learn
and from learning springs hope
and from hope springs action
to change our ways

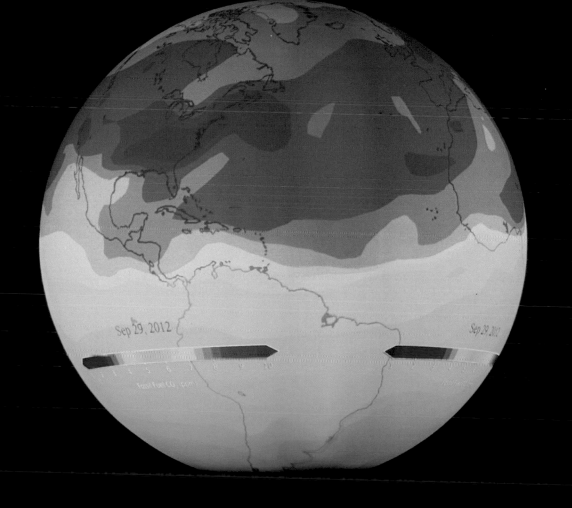

A view of fossil fuel emissions (oil, gas, coal), September 2012: Looking at the Western Hemisphere, the orange and red indicate a greater concentration of fossil fuels (parts per million). The green over South America is due in part to fewer people living south of the Equator.

maybe redemption comes as we remember
that one key reason we are alive
is to become better caretakers and stewards

by learning more about how Creation works
by respecting our space-traveling home
by using the resources we have been given
not just for ourselves but for everyone here today
and by not over-using the gifts of the Earth
so we can pass them on to all those coming tomorrow

by not focusing on me and mine
my family, my nation, my generation
but remembering that my family belongs to the human family
my nation is part of the global community of nations
my children are among all of God's children

maybe redemption happens
as we make our personal changes
not waiting for someone else to go first
and remembering
that we are all in this together
all 7.6 billion of us

maybe redemption comes from the hope we have
in knowing that we ARE changing some things for the good:
the great Antarctic ozone hole is shrinking
today's cars are coughing out 95% fewer fumes
than our grandparents' old Fords

Nature's way of managing carbon dioxide through the seasons: This view, from December 2005, shows more carbon dioxide being absorbed (green) in the southern hemisphere where it is summer, and less carbon dioxide being absorbed (white) in the northern hemisphere where the trees and shrubs have shed their leaves for winter.

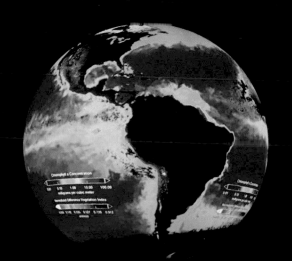

Marine chlorophyll provides nourishment to the world's fish, sea birds, marine mammals and humans. The green indicates areas of high concentration closer to the shorelines. The darker blue indicates lesser amounts of plant life farther out to sea.

Creation, Fall, Redemption

what our sacred texts offer is
is a pathway forward

Earth
    Erde
        Jorde
            Gaia

Planet Earth

we are all in this together
there's nowhere else to go

Canadian philosopher Marshall McLuhan spoke the truth:
"there are no passengers on Spaceship Earth
we are all crew"

and as crew
our part in Life's journey
is about freedom and responsibility
being responsible with our freedoms
and
becoming better stewards of

        God's gift of the Garden

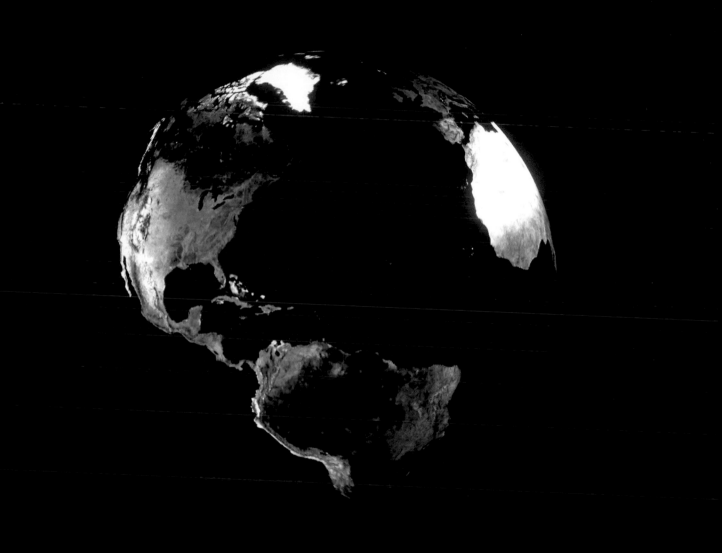

For the joy of human love,
brother, sister, parent, child,
friends on earth and friends above,
for all gentle thoughts and mild,

Lord of all, to thee we raise
this our hymn of grateful praise.

For each perfect gift of thine
to all people freely given,*
graces human and divine,
flow'rs of earth and buds of heav'n,

Lord of all, to thee we raise
this our joyful hymn of praise.

(last verse from John Rutter, b.1945,
*original: "to our race so freely given")

SUGGESTIONS FOR REFLECTION AND ACTION

* Ask yourself: which earth system do I think most needs attention today? Which would I like to learn more about (water, air, land, impact of human activity)?

* Consider: how might I use less of the resources that put carbon in our air and seas?

* Pass it along: share this book with a friend and have a conversation about what impressed you both.

* If you are involved in a religious community, ask your pastor, rabbi, imam, or leader to reflect on your tradition's story of Creation, and the work of being good stewards and Earth's caretakers.

* Visit the nearest Science on a Sphere location (see **sos.noaa.gov**).

* Recall the images and pictures of life on earth that bring you great joy and hope.

* Ask a science teacher how to calculate your carbon footprint.

* Invite a religious leader and a science teacher to give a presentation about how the ideas from religion and science can guide us to be more faithful and responsible.

* Decide which kind of trash is a serious problem to the land, water, and animals where you live, and invite your neighbors to address it with you.

* Give a copy of this book to someone.

* Say a prayer…each day…every day…thanking God for the whole Creation. And, write a prayer to share in your religious community.

* Visit a meeting of your local or regional political leaders and ask them to name the biggest earth-challenges in your community and what they are doing to address them.

* Investigate where people are improving the earth's ability to sustain human life and hold a party to celebrate.

* Organize a gathering of your neighbors to talk about ways you all could do something new to care for the Earth.

* Explore other environmental resources (books, groups, websites, films) and ask your friends to read, view and discuss them with you.

* Step back – close your eyes and "see" us always moving, spinning, revolving, traveling through the stars and galaxies.

* Step up – do something to help – as a person, a member of your family and community, as a citizen and as a human being riding our blue marble.

* Add your own ideas _____.

GRATEFUL ACKNOWLEDGEMENTS TO

* **Beth Russell**, the Science on a Sphere Operations Manager and supportive friend in this and several other SOS initiatives of which I have been a part;

* **Santa Fe Community College,** Santa Fe, New Mexico, host of one of the 150 SOS centers in the world and the only one to date in New Mexico, an institution that caught the vision of educating our younger generations in this most compelling way;

* Friends including **Ed Barker, Elaine Daniels, Harry Eberts, Vicki Fogel-Mykles, Dan Klein, David Maxwell, Dick Miller, Helen Maurose, Sheldon Sorge,** and **Ted Walkenhorst,** who read early drafts, made helpful comments, and improved the final manuscript;

* **Marney Wasserman**, pastor, author, wordsmith and my loving wife of 45 years.

ABOUT SCIENCE ON A SPHERE®

SOS is an outstanding resource being made available to our science centers, museums and educational institutions. There is a three-dimensional globe version (SOS) and a two-dimensional flat screen version (SOSx). Both have access to over 550 datasets available from NOAA. The SOS website provides information about both versions, the datasets, SOS locations, and how to acquire this resource. To learn more, go to: **sos.noaa.gov**

ABOUT THE AUTHOR

The Rev. Dr. Dave Wasserman is a minister in the Presbyterian Church (USA). When he retired in 2013, he visited his first SOS Center in Rockport, Texas. During his 40 years in formal ministry, he was actively involved in ecumenical and interfaith endeavors, initially serving congregations in Iowa, Arizona and Wisconsin, and later several districts (presbyteries) in Michigan, Oklahoma, Texas, and Arizona. He is not a scientist. He is one who appreciates the beauty, awe, and majesty of God's creation. The Wassermans live in northern New Mexico.

His other writings include: *Azure Wind (Lessons for Ministry from Under Sail)* reflections on his sailing sabbatical in 2006, and *Listen More, Laugh Often, Love Always (Reflections for Today's Church Councils)* written at his retirement moment in 2013.